SCIKU

STUDENTS OF THE CAMDEN SCHOOL FOR GIRLS

SCIKU

THE WONDER OF SCIENCE - IN HAIKU!

Edited by Simon Flynn
and Karen Scott

ICON

Published in the UK in 2014 by
Icon Books Ltd, Omnibus Business Centre,
39–41 North Road, London N7 9DP
email: info@iconbooks.com
www.iconbooks.com

Sold in the UK, Europe and Asia
by Faber & Faber Ltd, Bloomsbury House,
74–77 Great Russell Street,
London WC1B 3DA or their agents

Distributed in the UK, Europe and Asia
by TBS Ltd, TBS Distribution Centre, Colchester Road,
Frating Green, Colchester CO7 7DW

Distributed in the USA by
Consortium Book Sales & Distribution
34 13th Avenue NE, Suite 101
Minneapolis, MN 55413

Distributed in Australia and New Zealand
by Allen & Unwin Pty Ltd,
PO Box 8500, 83 Alexander Street,
Crows Nest, NSW 2065

Distributed in South Africa by
Jonathan Ball, Office B4, The District,
41 Sir Lowry Road, Woodstock 7925

Distributed in Canada by Publishers Group Canada,
76 Stafford Street, Unit 300
Toronto, Ontario M6J 2S1

ISBN: 978-184831-794-9

Typeset by Marie Doherty

Printed and bound in the UK
by Clays Ltd, St Ives plc

CONTENTS

FOREWORD

Writers have always drawn inspiration from science, perhaps most remarkably in poetry. Over two thousand years ago the Ancient Roman poet Lucretius sought to explain the world through observable physical principles. Two hundred years ago the Romantic Lord Byron wrote 'Darkness', a chilling verse portrayal of a world without sun, and in recent times the poet Lavinia Greenlaw has been in residence at both the Science Museum and the Royal Society of Medicine. Poetry and science have certainly developed an extraordinarily fruitful relationship over the years and it is a great pleasure to see students at The Camden School for Girls continuing that tradition of expressing in poetic terms the excitement and wonder of science.

As you dip into this extraordinary anthology, you will be amazed, amused and absorbed in equal measure. I hope that you will learn something new, but I am certain that you will feel great admiration for the students who have captured the thrill of their scientific and literary learning in such neat, economical use of language. I am very proud of all the students who contributed their work to a publication that will enable us to continue fund-raising for science and for the high quality of education generally in the school. I would like to thank Simon Flynn and Karen Scott for making this unique book happen. Finally my thanks go to the writers themselves, whose work will now encourage others to think poetically about science.

Elizabeth Kitcatt
Headteacher, The Camden School for Girls

INTRODUCTION

The world of science is an endlessly fascinating one, by turns intriguing, stimulating and thought-provoking. Yet all too often, when you ask adults about their experience of science at school, their response is a negative one: they were bored by it; they hated the practicals; they struggled with equations. Even some of those who have come to love it in later life didn't enjoy it at the time. Partly, of course, this is about the provision of school science a generation or two ago, didactic lessons taught with the aid of a blackboard and from behind a large and forbidding desk. And perhaps too it is about the nature of learning the foundations of science: establishing the basics of any subject is, after all, rarely as interesting as exploring the possibilities it later has to offer.

For many adults, though, their response to science echoes one of the dichotomies into which life can too easily be channelled: they were an arts person, not a science person; as a girl they didn't really do science – or perhaps they were a boy resistant to being pushed away from the arts and towards science, as was at one time all too commonplace. Mirroring C.P. Snow's idea of 'two cultures', we seem to pigeonhole ourselves as individuals into 'Science and Maths' or 'English and the Arts'. The beauty of the *Sciku* project is that it has combined both skill-sets, often seemingly polarised, enabling the pupils at The Camden School for Girls to play with language and literacy in order to communicate something scientifically complex concisely and imaginatively. This wasn't a forced or frivolous task. As Max Planck, the person who ushered in the quantum age, said, a pioneering scientist 'must have a vivid intuitive imagination, for new ideas are not generated by deduction, but by artistically creative imagination'.

Whether you're male or female, child or adult, someone with an arts degree or someone already interested in the subject, science is full of wonder, and in this series of haiku, we wanted to share the sheer fun and enjoyment of the subject with everyone, allowing you to learn something new or see it afresh in an unusual format. These haiku celebrate science in all its glory for readers everywhere and at every stage of their lives.

The act of adapting Japanese haiku, traditionally seventeen-syllable poems focusing on nature, into short poems about all things scientific – or 'sciku' – has in turn demonstrated the nature of our students: intelligent, inquiring, straight-talking, funny and at times downright bizarre. The range of poems that has been created in this book is testament to the enthusiasm the girls have for science and for expressing themselves confidently.

It began with the opening of a Year 10 chemistry lesson, when the girls were asked to write a haiku on polymers, which they'd learned about during the previous lesson. They embraced the task with real gusto and those haiku shared with the class proved to be witty, inventive and, most importantly, accurate in their science.

Next came a whole school science haiku competition, which ran during National Science & Engineering Week in March 2014. The winner (Jasmine Morris in Year 7) proved that poetry and physics can be combined to make something exciting:

A fallen apple
Eureka, it's gravity!
Shame about the bruise.

Finally, there is the book now in front of you. The students worked across the curriculum, spending time in both English and Science lessons, writing their poems and editing their work before submission. Each year group from Year 7 (aged eleven) to Year 12 (aged seventeen) was involved, and so the poetry spans a breadth of knowledge and abilities.

Our editorial team, made up of around twelve Year 9s and 10s (thirteen and fourteen year-olds), has been central in making decisions about this book. Giving up their lunchtimes to learn about the publishing process, discuss cover designs, select layouts and debate fonts, they have been lively and collaborative; keen to take ownership of the project.

It is vital that girls everywhere begin to see themselves as credible, erudite scientists who can go on to lead and be pioneering within science, technology, engineering and mathematics (STEM) careers. Even now it is still far too widely believed that girls can't – or just plain don't – do science as much as boys. This is a troublesome echo from a time when girls weren't afforded the same opportunities as boys, when society was convinced that science favoured the male mind. Instead, women like Caroline Herschel, Marie-Anne Paulze Lavoisier and Ada Lovelace were obliged to play a supporting role to their brothers, husbands and male colleagues. Even though their efforts frequently proved vital, they themselves were often invisible when the history books were written.

The gender imbalance within science has recently been in the news again, and perhaps it continues because of unconscious societal attitudes and even the toys and books that young boys and girls are given. Our technology department has this year run a project entitled 'Disrupting the pink aisle' in which students created designs for toys that would help foster and encourage girls as well as boys towards STEM. The feminist ethos at our school is strong, and we hope to play our part in changing the landscape of science by helping our girls to explore and feel confident.

The students at CSG are relentless at raising money for different charities. In 2013, four thirteen year-old Camden girls, part of the '6girlsnobuoys' team, swam the English Channel, breaking world records to raise funds for the school. We hope that the proceeds from this book will raise further, much-needed funds for the refurbishment of our ageing science labs (which date back to the 1960s) and help to give our students a better start in their science careers.

We hope too that you'll join with us in marvelling at the science here conveyed in delightfully succinct haikus, and that you'll have as much fun reading this book as the girls – and a few sixth form boys too! – had in producing it.

Simon Flynn
Karen Scott

Scikulogical barriers
To explain science
In seventeen syllables
Is really quite hard

SETTING THE SCENE

Science
It baffles your brain
But when you understand it
Satisfaction gained

The point of science
Science: created
To unravel the many
Mysteries of life.

A never-ending story
Science, what is it?
A pile of endless questions
Never to finish

What is it good for?
I think science means
A search for truthful answers
To stranger questions

Science vs. religion

Hot single dense point,
Big ginormous bang and that
Is how time began.

Seven days to make,
He has capability
To shape this wide world.

Spirit in the sky?
Or bang starting creation?
How did it begin?

It is argued that science and religion occupy
themselves with separate questions –
science deals with 'how' and religion aims
to answer 'why'. Of course, science would
argue that there is no need for the second.

Neutral science
Ever wondered if
Science is always what's good?
For ev'ry action

There is always an
Equal/opp'site reaction.
It is up to us.

Have faith
There is a fine line
Between true science and the
Act of believing.

Faith in science
Trust in the Big Bang.
Believe in see-through atoms.
Follow Newton's laws.

Science kills the mood
The love of your life?
A lump of quarks and protons;
Now there's a buzzkill.

John Keats, Edgar Allan Poe and Walt Whitman
are just three great poets who railed against
what they perceived to be the reductive nature
of science – but none of them made their point
quite as succinctly as the haiku above.

What is a hypothesis?
It's a prediction
Of what's likely to happen
That can be tested.

From geek to chic
Chemistry is cool.
Biology is awesome.
Physics is such fun.

Triple science
Atoms spin: Physics;
Birds and bees: Biology;
Test tubes: Chemistry.

HE-HE-HE

Funny fishy science

What kind of fish has
Two sodium particles?
It's a tu-Na fish!

The element of surprise

Oxygen went out
With magnesium today
I'm like OMg!

Okay, OMg is, strictly speaking, wrong – it should be MgO
because when naming compounds that include a metal,
the metal always comes first both in the name (magnesium
oxide) and the formula (MgO). But where's the fun in that?

He-He-He

Don't trust an atom.
Why not? Well, it's because they
Make up everything.

Making a meal of it ...
Humphry Davy, found
Six elements, shame he's dead.
We should barium.

Biological bash
A mushroom party,
Here comes the Portobello.
Hey! What a Fungi!

Talking of parties, there was one full of some big names in physics. Unfortunately, Pascal was under too much pressure to enjoy himself, while Einstein had a relatively good time and Ohm spent most of the time resisting Ampère's opinions on current events.

LOOK TO THE HEAVENS

Creation
Big, big, big, big, big
Big, big, big, big, big, big, big,
Big, big, big, big, bang.

Conundrum
The big bang theory
Not the show, the explosion.
What made it all start?

In the beginning ...
An immense bang cracked
Throughout the dark black abyss,
And so time began.

The universe is believed to be almost
14 billion years old. The Earth, by
comparison, is 4.5 billion.

Night sky
People wonder why
We can't touch the shiny stars
That sleep in the sky

I look at the stars
Our infinite universe
In all directions

They are there at night
They are there during the day
You just can't see them

While scientists generally agree that
the universe is flat, it's currently unclear
as to whether it's infinite or not.

Seirios
Sirius shining,
The brightest star in our skies;
Burning through our nights.

Seirios is Ancient Greek for 'scorcher'.

Astronomical art
Constellations, from
Humans to bears, joining stars
Like a dot-to-dot.

Stars
They live in the past
As huge burning balls of gas
Looking down on us.

Have you ever wondered why the sky is dark
at night, given that a star can exist at every
point in the sky? One suggested answer is
that the light from the great majority of stars
in the universe hasn't yet reached us.

Starry Potter
Bellatrix, Draco,
Lucius and Scorpius,
Why are stars evil?

Danger in space
Black holes are ruthless;
They suck in like a vacuum –
All things, even light.

The telescope
Beyond our vision
A galaxy of secrets
Suddenly revealed

Shooting stars
Light trails through the sky,
Meteoroid bright – burning
Millions tonight.

As pieces of debris from outer space (e.g. from asteroids or comets) enter the Earth's atmosphere, they heat up and emit energy due to the extreme friction and pressure they experience. This causes a 'fireball' or 'shooting star' to be seen across the sky. During this stage, the debris is called a 'meteoroid'; if it survives and lands on the Earth's surface, the resulting fragment is called a 'meteorite'.

Gravity in our solar system
Endless attraction,
A strong pull towards the Sun,
Cobweb of lovers.

Copernicus
We're not the centre
Copper knickers used to say
Now we know he's right

Sorry Aristotle
Earth, Mars, Jupiter
What do they have in common?
They orbit the Sun.

The Earth's orbit
Round and round and round,
Round and round and round and round,
Round and round the Sun.

24:7
One spin is a day
The four seasons come and go,
One orbit a year.

At the beginning of the 16th century, the general
consensus was that the Sun, and all the planets,
revolved around the Earth. In 1543, the Polish
astronomer Nicholas Copernicus published a book
proving otherwise. Copernicus' system had the
Sun at the centre and the Earth orbiting it, along
with all the other planets. This went against a
literal reading of the Bible and is one of the most
important moments in the history of ideas.

Burning bright
What is four thousand
Six hundred million years
Old but still so hot?

Solar radiation
A great ball of fire,
A huge surface explosion,
Our mammoth Sun flares.

Here comes the Sun
Halfway through its life,
Closest star to our planet
Burn you in seconds.

Oblivion
It will come someday.
Five billion years away,
The Sun won't make day.

Scientists currently believe that the Sun is about mid-way through its life and has a surface temperature of about 5,800K (kelvin). To put that into context, water boils at 373K.

The Goldilocks enigma
Too close we will burn,
And if it's too far we freeze.
The Sun is just right

Solar power
The Sun's energy
Drives the Earth's atmosphere like
A giant engine.

One area of major research at the moment is
how the Sun affects the Earth's atmosphere,
including whether Sun spots influence wind
patterns and whether cosmic rays seed clouds.

Big brother
Satellites in space
Watching your every movement
You better watch out!

The first artificial satellite, Sputnik, was
launched in 1957 and it's believed there are now
a few thousand in orbit around the Earth.

A hymn to Selene

Magical moonlight
High above me day and night
Reflecting the Sun.

Selene was the Greek goddess of the Moon.

Planet mnemonic

Order of planets:
My very eager mother
Just served us nachos.

Venus

The air is dead dense
On the cream planet of love;
Dangerous for you.

Red in tooth and claw

Mars, the red planet
Named after the great Roman
God of blood and war.

King of the gods
The fifth from the Sun,
Massive planet Jupiter,
Biggest of them all.

Satellites galore
Hermippe, Leda;
Two of Jupiter's many
Sixty-seven moons.

The last four planets
Jupiter, Saturn,
Uranus, Neptune ... all are
Immense gas giants.

A dwindling universe
Pluto: dwarf planet.
Now there are only eight left.
What if Earth vanished?

Calling all planets
Eight are recognised,
Pluto eliminated.
But are there more now?

Seventeen syllables –
the real reason why Pluto
is no longer a planet
Mercury, Venus
Earth, Mars, Jupiter, Saturn
Uranus, Neptune

Actually, there's an alternative reason for why Pluto was downgraded from a planet to a plutoid. During the 1990s and 2000s, it became increasingly clear to astronomers that Pluto was only one member of a diffuse collection of objects called the Kuiper belt, which lies beyond the eighth planet of the solar system, Neptune. This led to fierce debate regarding what constituted a planet, and that resulted in the following definition from the International Astronomical Union (IAU):

'A planet is a celestial body that
a) is in orbit around the Sun,
b) has sufficient mass for its self-gravity to overcome rigid forces so that it assumes a hydrostatic equilibrium [nearly round] shape and,
c) has cleared the neighbourhood around its orbit.'

Pluto fails to satisfy the final condition.

THE SPLENDOUR OF LIFE

The three ages of woman
First I'm a baby,
I'll have to be a lady,
Now I'm just a girl.

The splendour of life
The biology
Of every human being
Is incredible

Biology bake-off
Lungs, heart, liver, spleen.
The inner to my body,
Filling to my cake.

States of mind not matter
We are animals.
But with one key difference:
We have consciousness.

Being civilised
To be man or ape
It is our conscience only
Progress or regress

Grey matter
A place to store thoughts,
It soaks up things like a sponge.
Do we use it all?

The popular idea that we use only 10% of our brain is
a myth. Have a think about what that would mean.
Would you be the same if the other 90% were removed?
The truth is, we need, and use, 100% of our brain.

Cell-u-like

We are formed of cells,
Millions of these make us,
But only one me!

Not for eating

These animal cells,
They look just like small fried eggs;
'Can I eat them, Miss?'

Let's split up

What things do cells do?
They can do cell division.
Whoa, too many cells.

Cell specialisation

Correct stimulus
Some genes are switched on: active
Structure/role = altered.

White blood cells
Attack and engulf,
Fighting against illnesses.
Microbes have no chance.

We eat up microbes,
Producing antibodies,
Defusing toxins.

Immunity
Pathogens infect
Lymphocytes right to protect
They are our defence

The antibodies
Produced by our white blood cells
Protect our body

Inoculation
Using viruses
Now in the immune system
Forms antibodies.

Going viral
Pathogens mutate
Diseases created and
Epidemics spread.

Triple threat
Bacteria breed,
Viruses burst out of cells,
Toxins make us ill.

War games
Our white blood cells swarm
Foreign cells and digest them
Thanks immune system

White blood cells – my little soldiers
Cough, cough, cough, cough, cough.
Oh, I have caught a disease
My body is sad.

Pathogens please leave –
Viruses, bacteria.
White blood cells, save me!

Engulfing bad things,
White blood cells now save my life.
Hooray! I'm alive.

The immune system consists of white blood cells that
defend the body against pathogens. 'Pathogen' is a term for
any organism that causes disease. White blood cells carry
out their role either by 'engulfing' – therefore destroying
– the pathogen, by producing antibodies that inhibit the
pathogen or tag it (hence 'marking' it as a pathogen) for
nearby white blood cells to engulf, or by producing an
antitoxin that neutralises toxins released by the pathogen.

Get your facts right!
HIV and AIDS
Are not the same, but one can
Lead to the other

Friend or foe?

Some people might say
That statins are amazing.
I would not agree.

On paper they're great,
Lowering cholesterol,
That's all nice and good.

They help your body,
Inhibiting an enzyme.
Clearer arteries.

You might well now think,
It really is that simple;
You should think again.

They are quite evil,
Killing innocent people.
You should stay away.

> 'Statins' refers to a group of medicines
> that lower 'bad cholesterol', a key factor
> in heart disease, in the blood. There has
> recently been a lot of debate regarding the
> overall benefits, or otherwise, of statins.

P stands for phosphorus
Without phosphorus
Flowers wouldn't be healthy
And buds wouldn't show.

Phosphorus is needed by plants to convert light into
chemical energy during photosynthesis. It is important
for plant growth and the formation of flowers or seeds.

Soilsnake
Fleshy bits of string
Spreading out, engulfed in soil,
Writhing without eyes;

Freshening the earth
So baby plants can grow tall
And feed on nitrates.

Nitrates are needed by a plant in
order to make proteins. Its growth is
stunted if it doesn't get enough.

The fertility hormones
FSH hormone
Stimulates the ovaries
And an egg matures.

Oestrogen released
Inhibiting FSH
Starting another.

The hormone LH
Causes the release of eggs
From the ovary.

Three hormones and two glands are involved in the menstrual cycle. Follicle-stimulating hormone (FSH) is secreted by the pituitary gland in the brain. It acts on the ovaries (the second gland) in two ways: by prompting an egg to mature and by instigating the release of oestrogen. This second hormone also has two functions. It prevents further production of FSH and stimulates the pituitary gland to issue luteinising hormone (LH). This results in the mature egg being released alongside the increase in oestrogen.

Family planning
Oestrogen found in
Contraceptive pills prevents
FSH, LH.

If you miss a pill
Or you are sick or unwell
You might get pregnant.

Petal power
The flower is there
For flamboyant attraction;
Bees eat the nectar.

How bees pollinate
The plump bumblebee
Lands on the rose, collects the
Pollen and flies off.

He swoops over to
Another pretty rose where
The pollen is shared.

Fancy a drink?
Water drips down stem,
Picking up speed as it runs
Towards waiting roots.

The seven conditions of life
Mrs Gren stands for
Our seven processes like
Growth and nutrition.

Seven processes need to occur for something to be
defined as living. These are movement, respiration,
sensitivity, growth, reproduction, excretion and nutrition.
These can remembered with the natty mnemonic
'Mrs Gren' (also sometimes known as 'Mrs Nerg').

Ventilation
As we take a breath,
Oxygen flies to our lungs.
CO_2 whips out.

Staying alive
Heart keeps beating: it
Pumps oxygenated blood
Around your body.

The story of carbs
Plants store glucose as
Starch, which is insoluble,
For respiration.

The journey of an oxygen molecule
Breathing is so good:
Oxygen goes through your mouth,
Through the trachea,

Down through the bronchus,
Then the little bronchioles,
Through alveoli,

To pulmonary
Vein and the red pumping heart,
Travels round body.

Finally it's there,
And every cell we possess
Respires again.

Oxygen + glucose → water and ? (+ energy released)

The word equation
For respiration, includes
Carbon dioxide

Respiration can take place in living cells in two ways: aerobically and anaerobically. The haiku above refers to the first, which is when there is an adequate amount of oxygen. However, when this is in short supply, glucose is instead converted to lactic acid in animals and less energy is released. In plant cells, and certain micro-organisms such as yeast, glucose is instead converted into ethanol and carbon dioxide, and this is the basis of the manufacture of alcoholic drinks.

Show us what you're made of

Proteins in your cells
Do very important jobs:
Structure and enzymes.

Enzymes are complex molecules that control many of the reactions that take place in the body. For example, salivary amylase (that's right, it's found in your saliva) starts the process of digestion of starch. If you chew bread for a while, it should start to taste sweet. Biological washing powders contain enzymes that break down stains on your clothes.

Survival instinct
Animals need to
Move, find food, a mate, escape
From a predator

Food chain
Snails eat green leaves;
Birds eat snails, then they die;
Maggots eat the corpse.

A plain diet
Wildebeest eat grass;
Lions feast on wildebeest;
Hyenas scavenge.

A food chain shows what eats what, and
the first species in it is always a producer,
so-called because it makes its own food,
e.g. plants do this through photosynthesis.

Ebola emergency

Down in the Congo,
Passed on through fluids and blood
and there is no cure.

This sciku was written before the recent
outbreak gained attention in the British media.

Fitness

Heart pumping, breathing
Stretching muscles, gasp for breath
I'll now rest here, thanks

Food groups

Carbohydrates are
Potatoes, pasta, bread, rice
These are good for you

Protein we know is
Vital for growth and repair:
Chicken, eggs and pork

Fat can be evil
But is still crucial for us
In moderation

A recipe for a long life
You have to eat well.
Play sports, eat veg, don't do drugs
Don't smoke and don't drink!

The dangers of smoking
Stained teeth, smelly breath,
Mouth, throat, stomach, lung cancer
Asthma, heart disease

Class A advice
Some speed you up lots
And others slow you way down
Drugs: please don't take them.

Drugs are divided into a number of categories,
principally stimulants, depressants, hallucinogens,
opiates and performance-enhancing drugs.
Stimulants (e.g. crystal meth and nicotine) speed
up the central nervous system. Depressants
(e.g. alcohol and cannabis) do the opposite.

Not an E-asy Decision
Want to take a pill?
Unless it is in your will,
You should stay away.

Or, don't stay away,
Who cares what scientists say?
Your choice anyway.

Make no bones about it
Frame of our body
Stops us from being jelly.
It is very strong.

Bone of contention
There are two hundred
And six bones in your body.
I need all of mine.

We have an endoskeleton, which means that
it provides support from inside the body. An
exoskeleton provides support and protection
outside the body, and includes shells and the
outer structure of insects and crustaceans.

The bone connector
Ligaments are like
A tough band of elastic
Joining bone to bone.

Love me tendon
Tendon, a tissue
Connects muscles to our bones
Like a ligament

Antagonistic can be good
Muscle contracting
Extraordinary effort
Working in a pair

Many movements, such as bending our arms
or walking, require a pair of muscles in which
one contracts while the other relaxes. When
our biceps contract, our triceps relax. These
are called antagonistic pairs of muscles.

Musculoskeletal joy
Your bones and muscles
Create a system called the
Locomotors, yay!

Central nervous system
I decide to run.
My brain fires off messages;
I am on my way.

Chemical signals
From dendrites through to axons
Passing through neurons

Neurotransmitter
Diffuses across synapse
Carries the signals.

Good regulations
Homeostasis
Maintaining a steady state –
Not too hot or cold.

Body politic
We grow rapidly,
Sixteen to eighteen we stop
Due to genetics.

Short of breath
Emphysema means
That airways are narrowed and
Lung tissue breaks down.

Emphysema
Blocked, there's no escape,
Like cement filling the lungs.
Tar. Resist the urge.

Smoke-smothered child
Intoxication.
Lungs smothered in tar. Born to
Parents killing her.

What's the cause?
Is it genetic?
Or maybe deficiency?
They're types of disease.

Rickets (a lack of vitamin D), scurvy (a lack of
vitamin C) and kwashiorkor (lack of protein)
are all examples of deficiency diseases.

The gland scheme of things
Chemical message
That can make you go crazy!
Yup, they are hormones ...

Testosterone, oestrogen, insulin and
adrenaline are all examples of hormones.

A cell
Smallest living thing
Living, growing, dividing
Building blocks of life.

Cella [Latin: *small room*]
Nucleus: controls,
Cell membrane: Helps to protect;
Plants also have walls.

Animal and plant cells are similar in many ways but there are also some fundamental differences. Only the plant cell contains a rigid, cellulose wall. This is what gives plants their great strength. Only plant cells contain chloroplasts, structures that in turn contain chlorophyll. They give plants their green colour, and are responsible for photosynthesis.

Torn up beauty
An amaryllis
Ripped apart for a reason
To observe inside.

Many would argue that science is all about making observations and deriving our theories from them. However, as Sherlock Holmes critically put it, 'one begins to twist facts to suit theories, instead of theories to suit facts'. When a student looks at a dissected flower, it probably begins to make sense only once their teacher tells them *what* they're seeing.

The human cycle
Inside our body
Blood is pumping all around,
Until it goes cold ...

Microscope
Zooming further in,
The liquid we saw as pure
Now full of blotches.

It was the Dutch scientist Antonie van Leeuwenhoek who really put the microscope on the map in the 17th century when he discovered bacteria in plaque scraped from his teeth, as well as sperm and red blood cells.

The red army
Blood circulation,
Red and flowing, carries lots
Of haemoglobin.

Arteries and veins
Tubes carrying blood
Running through us head to toe.
To and from the heart.

Arteries take blood away from the heart
and veins take blood to the heart.

Blood
Sanguine, scarlet and
Contains haemoglobin, so
Oxygen's ample.

One of the most common misconceptions in science
is that blood can be blue. It's easy to see why the
belief is so widely held – our veins appear to be blue,
after all. If we deal with blood first, it is true that
only in the presence of oxygen is blood the vibrant
red we so commonly associate with it. But anyone
who has ever had a blood test will have realised that
the deoxygenated blood that is removed from their
vein isn't blue but is instead a deep maroon colour.
Our veins appear blue because of the way light is
absorbed and reflected by the skin and the vein.

Mitosis

Chromosomes condense,
Centrioles move to the poles
And produce spindles.

Chromosomes line up
As paired chromatids in the
Middle of the cell.

Spindles attach to
The chromosomes. They contract,
Pulling them apart.

Then the chromosomes
De-condense – identical
Nuclei are formed.

Mitosis is one of the two ways in which a cell
can reproduce (the other is meiosis). The key
difference is that in mitosis, an identical cell is
produced, whereas in meiosis the cells contain
half each of the original genetic material.

Maltose
Two glucose units
Form a glycosidic bond,
Releasing water.

Maltose is a sugar made of two
glucose molecules joined together,
with water lost in the process.

Solar power
The sun is shining
The leaves are absorbing light
Photosynthesis

Leaves of grass
Photosynthesis
Occurs in palisade cells,
Part of the process.

Plant solar power
Chloroplasts are green,
Absorbing Sun's energy
They are found in leaves.

Carbon dioxide
Two of oxygen
Plus a carbon atom makes
A gas that plants need

> Photosynthesis is the process in plants by which energy from the Sun is used to help produce glucose. This takes place in chloroplasts found in the palisade cells near the top surface of a plant's leaf. The equation for photosynthesis is the reverse of respiration described earlier:
> carbon dioxide + water $\rightarrow$ glucose + oxygen.

A skin full
The biggest organ
Is your skin; three layers, a
barrier to germs.

IQ test
World's greatest organ
Controls your every whim
What am I? The brain!

Digestion in a nutshell
Mouth, food pipe (gullet),
Stomach, then small through to large
Intestine, anus.

Break it down
During digestion
Enzymes will break down your food
Into useful parts

Enzymes are large proteins that speed up
reactions in the body. They're particularly
key in the digestion of food and include
amylase, which breaks starch into sugars,
protease, which breaks protein down into
amino acids, and lipase, which breaks
lipids (fats) into fatty acids and glycerol.

Bowel movement
The life processes,
Most amusing – excretion.
Through the rectum ... plop.

Autumn's gravity
Leaves fall from the tree
In different shapes, colours,
And crunch beneath you

Earth watch
Polar bears, panda,
Slowly becoming extinct
Keep them safe and sound

EVERYTHING UNDER THE SUN

Particles
We huddle in bricks
We dance around in water
We fly in the sky.

States of matter
Solids like to hug.
Liquids must be loose and free.
Gases like to fly.

Fluidity
Three states of matter:
Liquids and gases can flow
But solids cannot

Ancient elements
Earth, water, fire, air,
I'm sorry Aristotle,
What about carbon?

Everything is awesome!
Atoms are Lego
Building blocks, our universe,
Everything we see

The atom
From the furthest star
Through the universe it came
To make me and you

More than the sum of its parts
Atoms are petite,
They form bonds with each other
To make something new

Everything under the Sun
Atoms: everything.
Phone, shorts, water, food, laptop.
Also, they make you.

You are made up of
Those small, eeny, weeny dots.
They make all of us.

You cannot see them
With just the bare, naked eye,
But they surround us.

Joining atoms
Twos, threes, fours and fives
Countless combinations – so
Many molecules.

Double trouble
Joining together
Two or more types of atom
Gives you a compound.

A compound is a substance made up
of two or more different elements
chemically bonded together. A molecule
is a single, discrete unit of a substance
in which two atoms or more are held
together by covalent bonds. These bonds
can exist between non-metal elements
(e.g. water is H_2O) but not between a
metal and non-metal. It is possible for
a substance to be a compound and not
a molecule (e.g. NaCl, sodium chloride,
or salt) and vice versa (H_2, hydrogen).

Atomic orbits
Atoms have a small
Nucleus surrounded by
Many electrons

There is one exception to the above
– hydrogen has only one electron.

Inside the atom
Protons – positive,
Electrons are negative,
Neutrons are neutral.

Atomic optimism
Be like a proton!
Always positively charged
Never negative.

Atomic pessimism
I am negative
And repel others like me
Curse those electrons

Best friends really
The proton was sad
So, the electron said 'Smile!'
And be positive

An atom was walking along with his mate one day when he suddenly said to her, 'Drat, I've lost an electron.' 'Are you sure?' she replied. 'Yeah, I'm positive.' Atoms are composed of three sub-atomic particles: protons, neutrons and electrons. Protons have a positive charge and electrons are negative. Neutrons have no charge.

Homer Simpson's baseball team
Isotopes have the
Same numbers of protons, but
Different neutrons.

The number of protons an atom has tells us what element it is. For an atom to be a carbon atom, it must have six protons. Atoms have a neutral charge overall, which means the number of negatively-charged electrons must equal the number of positively-charged protons. However, atoms of the same element can have different numbers of neutrons and these are called isotopes. For example, there are three naturally occurring isotopes of carbon: carbon-12 (six neutrons), carbon-13 (seven neutrons) and carbon-14 (eight neutrons). In *The Simpsons*, Springfield Isotopes is the town's baseball team.

Elementary
Discovered and found,
One hundred and eighteen-ish,
Shown in a table.

A clear arrangement
Elements in rows,
And the elements in lines,
Make up a table.

Periodic table (1)
Many elements
Different bonding powers
Put into eight groups

Everything on Earth is made from one or more of the 118 chemical elements shown in the periodic table. The version of the table that we use today was first proposed in 1869 by the Russian chemist Dimitri Mendeleev, who wanted to illustrate recurring trends in the properties of the different elements. To do this he put them into a systematic order. This is what makes the periodic table such a powerful tool. Knowing this order allows us to predict how elements will behave, and how they will interact with each other or possibly combine to make new substances.

Building blocks
Carbon, helium,
Oxygen and hydrogen
Are all elements

Periodic table (2)
B, C, N, O, S
Half-metals and non-metals
That is chemistry.

Mercury
A liquid metal
An exception to the rule
And fatal if touched

S-elements
Scandium, sulfur,
Silicon, samarium
Strontium and tin!

Tin has the chemical symbol Sn!

Diamonds are for ever
The hardest substance,
A sweet structure of carbon,
Is a girl's best friend.

Diamond explained
Carbon is able
To form four covalent bonds
Beautiful di'mond

Allotropes
Di'mond and graphite
All made up of carbon but
None of them the same.

Allotropes are different structures of the same element. In diamond, each carbon atom is bonded to four other carbons in a rigid 3D structure; in graphite, each carbon is bonded to three others, resulting in very small layers of carbon that can slide over each other. As can be judged by the example above, their properties can be quite dissimilar – diamond is very hard whereas graphite is quite soft. Graphite can conduct electricity and diamond cannot.

Hero-killing elements
Six noble gases.
On the edge of the table,
One kills Superman.

Just add electricity ...
Amazing neon
Colourless, odourless gas
Yet it's a beauty.

The noble gases are extremely unreactive and, chemically speaking, are incredibly boring. However, when you pass an electric current through them, their charm is uncovered in a most spectacular way. Oh, and krypton and kryptonite are most definitely not the same thing.

A Curie-ous discovery
Radium is a
Hazard with high exposure
Silent, but deadly.

Radium was discovered by Marie Curie and her husband Pierre in 1898. It was initially used in luminescent paint until people realised the dangers.

Po
Tummy screens gone blank.
Polonium is fatal.
Oops, Dipsy is dead.

Just in case you don't get the reference
above, one of the Teletubbies is called
Po, which is also the chemical symbol
for polonium, a radioactive element.

Reactivity
Some things can react
Easily like sodium
Or not, like argon

The reactivity series part 1
Potassium and
Sodium react more than
Iron and silver.

The reactivity series part 2
Starting near the top:
Magnesium, zinc, copper,
Gold at the bottom.

Reactions
Pretty Sally could
Marry, but she decided
To combust instead.

The reactivity series is almost as powerful a tool for chemists as the periodic table. The series ranks metals in order of their reactivity and this helps chemists make predictions about whether certain reactions will take place or what would be needed to make them occur. Carbon and hydrogen are typically also included. Carbon is above iron in the series, which is why we're able to use it to extract iron from its oxide in a blast furnace (the carbon swaps places with the iron because it's more reactive, so we end up with carbon dioxide and pure iron). However, carbon is below aluminium in the series and so the same method can't be used to extract that metal from its ore. The most reactive elements are in group one, and a version of the series can be remembered with the mnemonic:

Pretty	(potassium)
Sally	(sodium)
Calmly	(calcium)
Married	(magnesium)
A	(aluminium)
Cute	(carbon)
Zombie	(zinc)
In	(iron)
Lovely	(lead)
Honolulu	(hydrogen)
Causing	(copper)
Many	(mercury)
Strange	(silver)
Gazes	(gold)
Perhaps	(platinum)

Group one
Group one elements
React quickly with water
Caesium explodes

Making a salt
Chips of tin bubble,
Reacting in pH 1.
Hydrogen and salt.

When a metal reacts with an acid (hence the pH 1), hydrogen gas is released and a salt is formed. For example, if the acid in the reaction above is hydrochloric acid, then the equation for the reaction would be:

$$Sn\ (s) + 2HCl\ (aq) \rightarrow SnCl_2\ (aq) + H_2\ (g)$$

(s) means solid; (aq) means dissolved in water; and (g) means gas.

The experiment
Fast, frothing acids.
Combine two in a beaker –
Please do not explode.

Chemistry in action

Fizz, bang, bubbles burst
Test tubes overflowing, fast
Fire burns, ablaze.

As exciting as the above sounds,
it's highly likely that 'questions
would be asked' of any teacher
who allowed it to happen in their
classroom. Health and safety, eh?

A real-life lesson

Scorching hot acid
Burnt off my eyebrow last year ...
But I learnt a lot.

Errm ...

Be careful!

Accidents happen
Science can be really fun
But not always safe.

A hazard
Ethanol: well known
'Highly flammable' substance.
Beware when using!

It's very important to be conscious of all potential hazards in a school science lab. With regard to chemicals, these tend to fall into the following categories: corrosive (e.g. concentrated sodium hydroxide and sulphuric acid), harmful (e.g. copper (II) sulphate), highly flammable (e.g. ethanol), irritant (e.g. dilute sodium hydroxide and sulphuric acid), oxidising (these increase the flammability of other substances, e.g. potassium manganate (VII)), and toxic (e.g. lead oxide). Hence no eating and drinking in a lab.

Risk assessment
Safety in the lab –
That is what we want to see.
Remember goggles.

Beware the burn
Acids, alkalis,
Are both very corrosive.
Careful when handling.

Ink spread
Never use pens with
Water. It will result in
Chromatography.

Acids and alkalis
On the pH scale
Alkalis go to fourteen;
Acids, nought to six

pH scale
A neutral substance.
Not acid or alkali
But in the middle

pH universal indicator
Judging by colour
Its nature exposed with hues,
Cherry, lime to mauve.

Green presents neutral
While red displays acidic
Blue shows alkali

pH (short for 'power of hydrogen') is a logarithmic scale indicating the concentration of hydrogen ions (H^+) in a solution – the lower the pH, the greater the concentration of hydrogen ions. The scale runs from 0 to 14 and is used to measure how acidic or alkaline a solution is. pH 7 is neutral and is the pH of pure water; above 7 and the substance is alkaline, below 7 and it's acidic. The further away from 7 it is, the more acidic or alkaline the substance. Just like the Richter scale that measures the strength of an earthquake and is also logarithmic, when the concentration of hydrogen ions changes by a factor of ten, the pH changes by only one unit.

Did you know?
Acid in your mouth,
And toothpaste on your toothbrush,
Make it all neutral.

Your teeth's enamel is affected at pH
5.5 (slightly acidic) and below. So you
might be surprised to hear the pHs of
the following foods and drinks: bananas
(5.1), black coffee (5), tomato juice (4.2),
orange juice (3.5) and cola (2.6).

Common a-salt
Acid, alkali
A salt and water are formed
Neutralisation

If you combine sodium hydroxide (caustic)
and hydrochloric acid (corrosive) in the
right proportions, you'll end up with
common salt (NaCl) dissolved in water.
This is called a neutralisation reaction.

A firework's colour
Strontium metal
Becomes crimson in a flame
Electrons falling

Lilac
The colour produced
When potassium is burned
In a roaring flame.

Metals, and their salts, have characteristic colours when burned. This happens because when the metal's salt is heated, the electrons in the metal jump up to a higher energy level. As they fall back down, the corresponding energy is emitted as light with particular frequencies that sometimes lie in the visible spectrum of light. This is exactly what happens in fireworks. For example, red will be produced by strontium or lithium salts; orange by sodium salts; green by barium; blue by copper and silver by magnesium or aluminium.

Fire
Its flame is so bright
We welcome the heat it brings
It gives us all light

The fire triangle
Fuel, heat, oxygen
The ingredients needed
For creating fire

Fire warning
Be careful if you
Have fuel, heat and oxygen
Fire can soon result

Complete combustion of fossil fuels
Excess oxygen
Hydrocarbons oxidised
No particulates

Buy yourselves an alarm
Carbon monoxide
Fills the air we breathe
Killing us swiftly.

Fires are the result of a chemical reaction between a fuel and oxygen. To get the reaction started, energy is needed, e.g. a spark or flame. The products of this reaction depend on the amount of oxygen present. Complete combustion of a hydrocarbon happens when there is more than enough oxygen available and this results in only two products: water and carbon dioxide. More often than not, however, not enough oxygen is accessible and we get incomplete combustion, where the products are water, carbon (soot particulates) and the poison carbon monoxide.

Gunpowder plot
Sulfur and charcoal,
Plus potassium nitrate
Means a great big bang.

Gunpowder was first invented
in China. Sulfur and charcoal are
the fuel, potassium nitrate is an
oxidiser and then all you need
to do is supply the energy.

Cold to the touch
An endothermic
Reaction absorbs heat, the
Temperature drops.

Reactions can be described as being either exothermic
or endothermic, and this refers to whether heat is
given out or taken in from the reaction's surroundings.
What this means is that exothermic reactions feel
warm (e.g. combustion reactions, and acid added to
a metal) and endothermic reactions feel cold (citric
acid added to baking soda, or eating sherbet).

Controlling a Bunsen burner
Covered hole: safety.
Medium: orange and blue.
Open: roaring flame.

The gas used in a Bunsen burner is mostly methane (CH_4), the simplest hydrocarbon. When you first light a Bunsen burner, the hole at the bottom of the burner's stem should be covered. This results in a yellow flame, which is also called the safety flame. As you open the hole, the flame turns from yellow to a roaring blue. The hole is affecting how much air, and therefore oxygen, is mixed with the methane before it's burned. This affects the level of combustion, which in turn affects the amount of energy produced in the chemical reaction that results in the flame. The burner must always be set at the safety flame when not being used to heat apparatus.

Sulfur dioxide
Pollution rises
Rain returns it to the Earth
This is acid rain

Sulfur, oxygen
Then water make an acid.
Don't encourage it.

Crude oil, a fossil fuel, is a mixture of hydrocarbons and can be separated into its different components (e.g. petrol and bitumen) using fractional distillation. Hydrocarbons are compounds containing only carbon and hydrogen. However, sulfur impurities also exist in fossil fuels. This means that sulfur dioxide is also released into the atmosphere, which can subsequently lead to the formation of sulfuric acid.

A literary poison
Cyanide is great
For killing lots of people
In detective books

Any chemical compound that contains a carbon atom triple-bonded to a nitrogen atom is classed as a cyanide. This includes substances in certain plants such as cassava, lima beans and almonds. It is also a form of defence in some millipedes and centipedes. So, any fictional murderer has a wide range of sources from which to make their poison of choice. Hydrogen cyanide, which is said to smell like bitter almonds, kills by preventing the body's cells from using oxygen properly. This, in turn, stops the process of respiration.

Wow!
Much stability
Such unreactivity
Gas nobility

A noble delight
Helium: a gas
Fills little children's balloons
Makes voices higher.

A noble caution
Helium's a gas
It makes your voice go high-pitched,
May go light-headed.

Helium is the second element on the periodic table and is a noble gas. Like its stable-mates, it is very unreactive. This is one reason why it was first discovered through spectral analysis of a solar eclipse rather than being isolated on Earth (that came nearly 30 years later). It is named after Helios, the Greek god of the Sun.

Dissolving confusion
To some, solutions
Are answers; to chemists they
Are still all mixed up.

In chemistry, a solution is a solute (e.g. sugar) dissolved in a solvent (e.g. water) and is indeed a mixture. The sciku above goes very nicely with an old chemistry joke: 'If you're not part of the solution, you're part of the precipitate.'

Liquid of life
Water is made of
Oxygen and hydrogen,
Which we need to live.

Water, water, everywhere ...
Oh, how I wish we
Did not study H_2O,
Just drank it instead.

Drink it all in
Water may seem dull
But is a really vital
Substance in our lives.

Water
We should cherish it
Our survival, just one life
Crucial to conserve

Water is essential for life because it's the solvent in which all metabolic reactions occur – nothing else would work as well. Between 55% and 60% of an adult's body is water.

The test for an alkene
Add bromine water
To a test tube of alkene
The colour is lost

Polymerisation
Double-bonds broken
When alkene and alkene meet
Forming polymers.

An alkene is a hydrocarbon with the general formula C_nH_{2n}, for example ethane from which polythene is made. Polymers are large molecules made up of thousands of repeating units. Examples of natural polymers include starch, silk and DNA. The first synthetic polymer, Bakelite, was invented in 1907 and heralded the dawn of the plastic age. Other modern polymers include nylon, Kevlar and Teflon.

Oxidation of alkenes

Alkenes' products of
Oxidation are diols
With two OH groups.

In these reactions,
Dilute sulphuric acid
Is necessary

With potassium
Manganate VII or else
With dichromate VI.

And respectively,
Purple becomes colourless
And orange turns green.

Hydrogenation

An alkene reacts
With H_2, forming alkane.
Catalyst: Nickel.

Small changes ... big effects
Sudafed. Crystal
Meth. Only one OH group
Separates the two.

Incredible as it may seem, a type of nasal decongestant and the class A drug crystal meth are structurally identical except where the former has an OH group (an oxygen and hydrogen atom joined together) attached to it, the latter has just an H (hydrogen). In the TV series *Breaking Bad*, the first batch of crystal meth Walter White produces begins with an over-the-counter drug.

Electronegativity

The ability
To attract electrons in
A covalent bond

Makes the difference
Between an ionic bond
And a covalent.

Periodic trends
Show an increase to right
Decrease going down.

Fluorine has the strength
Much more than any other.
Francium the least.

Pauling discovered,
Back in nineteen thirty-two,
The trend and his scale.

Students don't come across electronegativity until they study A-level chemistry. It was first proposed by Linus Pauling, one of only two people to win a Nobel prize in two separate fields (the other was Marie Curie). Electronegativity refers to the ability of an atom to attract a pair of electrons in a bond. If two bonded atoms have the same electronegativity, the pair of electrons will be on average found equidistant along the bond. However, when there is a difference in electronegativity, the electron pair will be on average closer to the more electronegative atom. This can have a big effect on a molecule's reactivity.

Volumetric analysis
Pipettes and burettes
Hold acids and alkalis
During titration.

Titrations
Oh woe betide you
For any funnel left in
One of your burettes.

A titration is a method of analysis in chemistry that allows you to calculate the concentration of a substance, such as an alkali, by using a known concentration of another substance that the first reacts with, such as an acid. A burette is like a very long measuring cylinder with a tap at one end that allows you to deliver what's in the burette very accurately, typically +/– 0.05cm^3. However, this accuracy can be compromised in various ways, including leaving the funnel you used to add what's in the burette at the top. This is because drops can still trickle down from the funnel and add to what you thought you had. A volume pipette allows you to measure a particular volume, e.g. 20cm^3, very specifically.

LET'S GET PHYSIC-AL

Supporting physics

Physics and its laws;
The bones of our universe,
Like the bones in me.

Let's get physic-al

Electric currents,
Electromagnetic waves,
Energy, forces

Archimedes' discovery
Volume of body
Equals volume of water
Displaced – Eureka!

The story goes that Hiero, the king of Syracuse where Archimedes lived, commissioned a golden crown to be made using a precise amount of gold. This was duly delivered and the king was very happy until it was suggested that some of the original gold had been replaced with cheaper silver. Hiero needed to know for sure if he'd been duped, and called for Archimedes' help. This required some thinking. As Archimedes was lowering himself into a bath and seeing the water level rise, a way of knowing for sure whether the crown was pure gold came to him and he leaped out of the bath and ran home naked, shouting 'Eureka!' – 'I've found it!' Archimedes collected masses of gold and silver, each with the same weight as the crown. However, because gold and silver have different densities (density is mass/volume), the same masses of each have different volumes. Archimedes would be able to measure these volumes by seeing how much the water level rose when each mass was lowered into the liquid. He did this with a mass of gold, then a mass of silver, and then the golden crown – and discovered that the crown was indeed a mixture of gold and silver. The crime was revealed.

Particle theory – solids, liquids and gases
Fixed but vibrating
Touching but able to move
Flying here and there

In a state
Is sand a liquid?
Sand has solid particles
Like a broken rock.

The question asked above is not as strange as it may seem. Scientists have discovered that if sand is allowed to slide down a rough slope, it behaves in ways similar to solids, liquids and gases. Amazing!

From state to state
Condensing downwards,
Plip! Plop! Puddles start to form;
Evaporates. Gone.

All spread out
High to low, high to
Low, the particles diffuse,
Spreading round the room.

I think the above may be a polite way of saying my after-shave may be too strong (scents reach us thanks, or no thanks, to diffusion through the air). Diffusion can also take place in a liquid, where particles are free to move. It's because of the random motion of particles that diffusion happens. For example, imagine if you put a load of folded clothes in a dryer and then ran it for twenty minutes – how likely do you think it would be for you to open the dryer and find your clothes refolded? Much more likely would be that they were all messed up.

Conservation in physics
It's only passed on
From grass to sheep to humans.
It's not created.

According to the first law of thermodynamics, 'energy can be neither created nor destroyed, only interconverted between stores'. This law is in effect based on the principle of conservation of energy and is one of the most important laws in science.

Energy
Up, down, all around
Energy is to be found
But you can't see it

True colours exposed
Bit of a shocker,
White light is a colour mix,
Creating rainbows.

A wonderful palette
The purest white light
Prism divides an array
Of beautiful shades.

Using a prism
Entering as one
White light splits into seven
Visible spectrum

Light through a prism
Red, orange, yellow,
Green, blue, indigo, violet;
Pink Floyd had a point.

There's more to Newton than apples you know
Isaac Newton liked
To observe the colours and
How human eyes worked

When Isaac Newton first conducted his experiment separating white light using a prism, he initially wrote of there being only five colours – red, yellow, green, blue and violet. He later added orange and indigo, partly because he was eager to show a connection between colour and music – the seven colours matching the seven notes of the ancient Greek Dorian scale. Visible light is, in fact, a continuous spectrum, which Newton himself was well aware of.

Spectrum
Light shines through raindrops;
Wavelengths split, spectrum colours,
Rainbows dance through life.

Raindrops act a little like tiny prisms — they refract (or 'bend') and reflect light as it passes through. Each wavelength of light refracts different amounts: violet light refracts the most, while red light refracts the least — producing a spectrum of colours.

How we see this poem
We see things because
Light rays reflect off objects
Into our pupils.

The colour of an object depends on which parts of the visible spectrum are absorbed and which are reflected. For example, leaves are green because they absorb red and blue light and reflect green.

Beyond the rainbow
All kinds of objects
Emit AND absorb infra-
red radiation

First discovered by the astronomer William Herschel, infrared light is just beyond the red end of the visible spectrum.

Colour
The green I see could
Be a completely diff'rent
Colour t'what you see

Seeing is believing
Light is a spectrum
From violet to crimson,
Only in our eyes.

Light
Travels in straight lines
Much, much, much, faster than sound.
Too bright for our eyes.

The speed of light is almost 3×10^8 metres per second, whereas the speed of sound in air is around 340 metres per second. This explains why it's possible to estimate how far away lightning is. For every second after you see a flash of lightning, the sound produced travels 340 metres. So, if you hear thunder three seconds after seeing lightning, then it must have been 3×340 metres away or approximately 1 km.

Forces
Contact, non-contact.
Isaac's three laws of motion.
Balanced, unbalanced.

Forces in action
Magnets pull metal
Divers swim underwater
Cars can drive forward.

Newton's first law
Forces are balanced
Equilibrium is reached
Object is at rest

or

Forces are balanced
Equilibrium is reached
Uniform motion

A body will stay in a state of rest or will continue
to keep moving in a straight line and at a constant
speed unless an unbalanced force is applied.

Friction
I walked down the street
And slipped on a banana
Reduced resistance

Friction is a resistant force that comes about when
solids and/or fluids (a liquid or gas) come into
contact with each other. Air and water resistance are
examples of friction between a solid and a fluid.

Resistance is NOT futile
Friction gives you grip,
Friction can cause you to stop,
Friction creates heat.

Is work difficult?
Further from pivots
Levers make work easier
As they apply force

In physics, 'work' is defined as the displacement of an object by a force. For example, in a waterfall, gravity (a force) acts on water, causing it to be displaced downwards (to fall). When a lever is used around a pivot, the force needed to move an object is reduced, and the further from the pivot the force is applied, the less force is needed (think of the difference between using a nail and a screwdriver when trying to open the lid of a paint tin). Archimedes summed it up nicely when he said: 'Give me a place to stand and a lever and I will move the Earth.'

Under pressure
Working out pressure
To do this, you divide force
By the area

The unit of pressure
is the Pascal (Pa).

Newton thinking
A fallen apple
Eureka, it's gravity!
Shame about the bruise.

Gravity
An attractive force
Between all objects with mass
Just like you and me.

According to Newton's law of universal
gravitation, a force of attraction exists between
any two or more objects with mass. The
strength of this force depends on the size of
the masses and the distance between them.

Gravity on Earth
The force that helps us
To keep our feet on the ground,
Otherwise we'd drift.

Flying high
Gravity, a force
Pulling me down to Earth though
My head is in space.

Watch your step
Don't put gravity
To the test or you will fail;
Fall and it will win.

Flight
Aerodynamics,
'Like Icarus ascending
On wild, foolish arms.'

Use the force
I jump from the plane,
I release my parachute,
Air resistance saves.

Have a safe flight!
The parachute falls,
Air resistance strikes, pushing
Up, gravity down.

Crying
Saline drop reaches
Terminal velocity
Then, obliterates

A science lesson from Yoda
Iron, cobalt and
Nickel are the magnetic
Elements – these are.

It's simple
Opposites attract ·
And like repels like, that is
Magnets, don't you know.

Non-fatal attraction
Moving the magnet,
Iron filings gaily dance,
Too close, all sucked up.

As the magnet moves,
It's like follow the leader,
Iron soldiers dance.

North and south
Molten iron core
Makes a strong magnetic field
That surrounds the Earth.

At the centre of the Earth is a solid inner
core surrounded by a molten outer core.
The former is made mostly of iron and
the latter iron and nickel. Convection
currents in the outer core, coupled with
the spin of the Earth, result in a magnetic
field being formed that surrounds
the planet. Technically speaking, the
Earth's magnetic north pole is really
the south pole of the magnetic field.

Electricity
Powering our world
We would be lost without it.
Everything would stop.

She's electric
I run round circuits
Transfer energy from cells
Light up any lights.

Electric circuits
Flowing electrons,
Must be negatively charged.
Round and round they go.

Electrons carry the charge around a circuit. Interestingly enough, 'conventional current' considers the current to leave the positive end of a power source and travel towards the negative end. In fact, as electrons are negatively charged, the actual current travels from the negative to the positive end of the power source — in the opposite direction to how we conventionally see it.

Scientific Russian dolls
A tiny atom,
Its nucleus is minute.
Quarks are much smaller.

The end of the road ...
Quarks and leptons are
The only fundamental
Particles ever

Quarks and leptons are what make up matter. Quarks come in six 'flavours': up, down, charm, strange, top and bottom. Electrons are examples of leptons, of which there are six flavours: electron, electron neutrino, muon, muon neutrino, tauon and tauon neutrino. The electron, muon and tauon all have a charge of −1 and differ only in mass (electrons are smaller than muons, which are in turn smaller than tauons). Neutrinos have no charge.

Hot stuff
Heat can be transferred
By conduction, convection
And radiation.

Pardon?
It travels in waves.
A shout, bell or car alarm.
All sound acts this way.

Music to our ears
Vibrations ripple
We hear the sound – loud or soft
All from little hairs!

Longitudinal ...
Waves run parallel
To the direction of the
Energy transfer.

All waves involve vibrations. However, there are two types of wave motion – longitudinal and transverse. Water and light waves are transverse, which means their oscillations (vibrations) are at right angles to the direction the wave is travelling. Longitudinal waves, such as sound, have vibrations in the same direction as the wave is moving. If that doesn't make sense, imagine a stretched out slinky being pushed at one end and watching that compression travel to the other end. We hear sounds because these vibrations are picked up by tiny hair cells in our ear. The larger the vibration (its amplitude) is, the louder the sound. And the more vibrations reaching the ear per second (its frequency), the higher the pitch.

Domestic science
Schrödinger's pet cat
Which is he? Alive or dead?
Poor Mr Whiskers

Schrödinger's cat is one of the most famous thought experiments in science and was designed to draw attention to one of the strangest elements of Niels Bohr's version of quantum theory, which was that sub-atomic events existed as probabilities that were resolved only when the event was observed in some way. In Schrödinger's thought experiment, a cat is placed in a steel container with a tiny bit of radioactive material. Over the next hour, there is an equal chance that an atom in this material will decay or not. If it does, a poison will be released and the cat will die. The point was that the cat's state would be resolved only when someone peered into the container. Until then, the cat would be both alive and dead. To Schrödinger, this was absurd.

Time travel
What would happen if
We moved at the speed of light?
Whoops, it's 3012.

According to Einstein's special theory of relativity, it's possible to time-travel far into the future, provided that speeds close to that of light are reached. However, it rules out the possibility of ever coming back from the future.

THE RHYTHM OF LIFE

Creationist or evolutionist?
Old Adam and Eve,
We are all being deceived.
You should trust Darwin.

The evolution of cool
Darwin once chose to
Shave his head, never his beard.
The first hipster born.

The rhythm of life
From a fish onwards,
From an ape to a human:
From cradle to grave.

Evolution
Billions of years
To bring us to this moment
Time still ticking by.

No more hair suits
Do humans evolve?
Yes, we do change over time
Goodbye body hair

Survival of the fittest
Evolution is
Surviving the big changes
Thanks, adaptation

Adaptation
Variation makes
A change to a body to
Fit a location.

Polar bears adapt
Small ears, milk-white fur
With sharp teeth and strong legs but
Can they face no ice?

Desert survival
Camels can adapt.
They use their humps to store fat.
Well, how about that?

One widely-held misconception in our understanding of evolution is that it's teleological, i.e. that it is directed, has a purpose or aim. It does not! Even Radio 4's *Today* programme has been known to fall into this trap, for example when it discussed how the bacterial infection MRSA was 'finding ways' (this implies conscious thought, doesn't it?) of becoming resistant to antibiotics. The truth is that evolution is completely random. Variations within species come about by chance but once present can become an advantage. If that advantage proves to be the difference between survival and death, then it will become increasingly common because those organisms possessing it will survive long enough to reproduce. The phrase 'survival of the fittest' means that the organisms best adapted to the environment they live in will be the ones most likely to survive and have offspring. ('Fittest' means 'best fitted' and not 'physically the most fit'.)

Home sweet home
Safe in habitats.
Don't wander to another.
Survive in your own.

Camouflage
Peppered moths were light,
Then dark and now light again.
Nat'ral selection.

The fortune of the peppered moth over the last 300 years is often held as a classic example of natural selection in action. The moth typically comes in one of two shades, light or dark. Before the industrial revolution in the mid-18th century, the lighter-shade moth had the advantage in survival because it was better camouflaged against the more common light-coloured trees. However, when factories started pumping out great quantities of soot, the average shade of trees grew darker and the light-coloured moth became much more vulnerable, dramatically decreasing in number. During the 20th century, air quality improved and trees became lighter again, heralding the return of the light-coloured moth.

Conception
Sperms look like tadpoles,
Their head breaks into the egg,
It breaks to protect.

A nine-month journey
First an embryo
Then a small growing foetus
Leads to a baby

Foetus
Pressed inside tight space,
Within nine months of splitting,
Cells become a life.

New life
Birth is wonderful
Children get their parents' genes
And keep them. Always.

Blame it on your parents
You have your mum's genes,
And your dad's as well, you know?
We have forty-six.

Strand-ed
You *are* DNA,
Everyone *has* DNA,
DNA *is* you.

The instruction manual
A genetic code,
Storing inside your body,
Decides how you look.

A rich tapestry
Blue eyes and brown hair,
A DNA patchwork quilt,
Woven into life.

In both plants and animals, DNA is packaged up (along with other materials) in the form of chromosomes, which look like two thread-like structures. There are 46 chromosomes in 23 pairs in every cell of the human body. The disorder Down's syndrome is due to a person having an extra copy of chromosome 21.

All it takes ...
A nucleotide
Contains just three components:
Base, sugar, phosphate.

DNA is a natural polymer made of repeating nucleotides, each of which contains a phosphate group, a deoxyribose sugar and one of four nitrogen-containing bases: adenine, cytosine, guanine and thymine. The bases are given single-letter abbreviations (A, C, G, T) and it's from these that we grasp genetic code. The dystopian film *GATTACA*'s title comes from these letters.

That's stretching it ...
DNA can stretch
To the Moon and far beyond
It codes our bodies

If you unravelled the DNA from just one human cell, it would be approximately two metres in length. If you then did the same for every cell in the human body and laid each strand end to end, it would roughly reach the edge of the solar system.

A question of chromosomes
Mum, are you XX?
Dad XY? Assemble them
And that makes me, right?

Battle of the sexes
Sperm reaches the egg,
X and X, a baby girl!
Genes decide gender.

One of the 23 pairs will be made of two X chromosomes if you're female and an X and Y chromosome if you're male (only men carry a Y chromosome). When sperm and egg cells are formed, only one chromosome from each pair ends up being in each sperm or egg cell. So, in every egg cell there will be an X chromosome (either one from the female's XX pair) whereas half of the sperm cells will contain an X chromosome and the other half a Y chromosome (either one from the male's XY pair). When a sperm and egg cell fuse, the number of chromosomes is restored to 46, with half of the genetic material coming from the father and the other half from the mother. Whether the sperm cell involved contains an X or a Y chromosome will ultimately determine the baby's sex.

Pseudoscience part 1
Five to ten per cent
Shows that when you are young, you
Have attractive kids.

Pseudoscience part 2
Choose a young partner;
Old men give ugly babies;
Never get with them.

Pseudoscience part 3
Old men's genetics;
Scaremongering mothers. Why?
Seems dystopian.

The above haiku were written in response to articles in the British press reporting the findings of an anthropologist at Vienna University that 'someone born to a father of 22 is already 5%–10% more attractive than a 40-year-old father'. Some saw the reporting of these findings as a way of putting pressure on women to have children younger. Hmm …

Heredity
Like father, like son
Diseases can be passed on
Through generations.

Allele together now
Twenty-three pairs of
Chromosomes; some can be seen,
Some hide in your genes.

Alleles are different versions of the
same gene, e.g. the gene for eye
colour could be blue or brown. Genes
are found in chromosomes and
because these come in pairs (one
from each parent), everyone also
has pairs of genes. The next sciku
explain what effects might result ...

Gene genie
Blue eyes or brown eyes?
Which is the dominant gene?
It has to be brown.

Traits such as eye colour can be said to be
either dominant or recessive. The dominant
will always mask the recessive. This explains
particular patterns of inheritance and
why traits can skip generations. The gene
for brown eyes is dominant over green,
which in turn is dominant over blue.

Drawing the short straw
My growth cells are done,
But I'm staying a midget.
We've tried everything.

From the inception of life, the codified information in DNA affects the way an organism grows and what it becomes.

Body imaging
I like my body
No one can tell me I'm fat
It is my body

Protein synthesis
A burst of enzymes:
A transcription traffic jam.
Watch for gene speed-bumps.

Franken-science

Soon, in the future,
We'll be able to change a
Baby's DNA

Cloning v1

Egg cell stripped apart
The nucleus cast aside
Empty cell awaits ...

Skin cell stripped apart
The nucleus is exposed
Awaits new shelter

The egg cell is filled
It shelters the nucleus
A fusion takes place

It splits again and
Again, multiplying. More
Embryos are formed

The host takes its place
Embryo is inserted
A creature is born

Cloning v2
Embryo transplants
Genetically engineered
Adult cell cloning

Body cell taken
From an adult female sheep
Nucleus removed

An egg cell taken
From the second female sheep
Nucleus destroyed

Fused cell develops
Then placed in uterus of
Surrogate mother.

The infamous Dolly the sheep was the first mammal cloned from an adult cell, via transfer of the nucleus. Dolly had three mothers: the first provided the DNA (which was taken out of an adult cell from the udder — this made the sheep Dolly's biological mother), the second provided the egg cell (in which the original DNA was taken out or 'emptied' and the new DNA inserted), and the third was the surrogate mother, who carried the implanted embryo until birth. This was such an important breakthrough because it showed that the genes in an adult cell could still return to the state of an embryo, which could give rise to a full mammal. Dolly lived for 6½ years, almost half the average for her breed. Since 1996, many large animals have been successfully cloned, including bulls, horses and a rhesus monkey. Natural clones do occur, most notably in identical twins.

Lysogeny
Virus insertion,
Dormant virus in a cell,
Prevents transcription.

There are two life cycles that a virus can have – lysogeny is one of them. Here, a virus inserts its genetic material into a cell and this is assimilated into the cell's chromosome. However, instead of becoming active immediately (to make new virus particles), its genetic material becomes dormant until a time of stress on the cell and then it activates, for example in the case of cold sores. When the cell replicates, the virus DNA is replicated also.

THE BIG PICTURE

Gender bias

Is the research true?
Women get better results
But men control labs.

Proportional representation

Something you should know
Yes, reject the status quo
Science needs women

Girls typically outperform boys in GCSE exams, including the sciences. Yet this doesn't translate into healthy numbers of girls taking science subjects further. A study by the Institute of Physics in 2012 found that nearly half of all state schools did not have any girls doing A-level physics. Furthermore, WISE (Women in Science and Engineering) reported that only 13% of all STEM (science, technology, engineering and mathematics) jobs in the UK are occupied by women.

Credit where credit is due

Franklin – DNA
It's a shame she died so young.
Nobel prize denied.

The scientist Rosalind Franklin played
a pivotal role in the discovery of the
structure of DNA. Sadly, she died in 1958
at the age of 37 and, as Nobel prizes are
not awarded posthumously, she was
ignored when Crick, Watson and Wilkins
were awarded theirs four years later.

Our shared planet

To be man or ant,
We all voyage together,
In this befouled world.

After life

When we are buried
Each body disintegrates
Enriching the soil.

650 million years later ...
Some past animals
Become fossils while others
Were turned into oil.

Natural recycling
Squirrels get eaten
And gaudy flowers get picked
Microbes start their feast.

Process of decay
Releases CO_2 in
To the atmosphere.

The carbon cycle: a modern epic
You know, carbon is
In our huge atmosphere as
Carbon dioxide.

Photosynthesis
Is where carbon dioxide
Is absorbed by plants.

Carbon compounds are
Passed from plants to animals
By being consumed.

O respire my plants!
O respire my animals!
Return CO_2!

Carbon, the basis of all life on Earth, is constantly being recycled. Both terrestrial and oceanic plant life take carbon dioxide in from the atmosphere and the carbon becomes stored in their living matter, which may then be consumed by other living organisms. Some of this carbon is returned to the atmosphere via respiration. When living organisms die, they are eaten by decomposers, which also respire and so more CO_2 is given out. Some organic material avoids decomposition and, over a long period of time, is turned into fossil fuels. When this and any other organic material is burned, CO_2 is released.

A plea for renewable energy
What are we to do
When our fossil fuels run out?
Nuclear scares me.

Rising temperature and sea levels
Increase the heat and
The Arctic and Antarctic
Melt into the sea.

Climate change
When the icecaps melt
And the land begins to shrink
What do we do next?

Global warming
Cars and people make
Greenhouse gases in our world.
Walk to school, don't drive!

Help the Arctic
We burn fossil fuels.
Ice caps melt; the Earth heats up.
Polar bears homeless.

Global disasters
Death by natural cause,
Devastating floods and droughts,
Caused by CO_2

The killing
CO_2 released,
Trapped in a heating oven.
Mother Earth's murder.

The reality of climate change
Ice sheets in retreat
'The tipping point has been passed'
Time is running out.

'It's a false alarm'
Blinded by such contention
Our planet dissolves.

Tundras dissipate
Kinetic energy soars
Act coherently

Earth's slow destruction
'It's a happening now thing'
Too late to go back

Wake up, can't you see?
Reality has arrived
Death is upon us.

Our increasing reliance on fossil fuels since the 18th century has resulted in more CO_2 being released into the atmosphere during that period than has been taken in by plant life. CO_2 is one of many greenhouse gases, all of which act to reduce the amount of heat that escapes into space from the Earth. An increase in the concentration of greenhouse gases results in an increase in the average global temperature that is having a huge impact on life on Earth.

Last
Last fossil fuel burned,
last greenhouse gas relinquished,
The last breath we take.

A goodbye message
Ice caps are melting.
As the temperature rises;
Farewell planet Earth.

Dangers of ocean acidity
Extinction proceeds,
As water acidifies.
Creatures, they die out.

Another consequence of the rise in CO_2 in the atmosphere is an increase in the acidity of oceans. This happens because, when oceans absorb CO_2, carbonic acid is formed. This has a knock-on effect on marine life. Coral reefs, in particular, are sensitive to the pH of the water they live in.

Total wipe out
Extinction is a
Natural process of life
But it's too fast now.

Extinction of animals
The future is death,
We are destroying the world.
Silent assassins.

Each year species die;
This is due to our actions.
Yet, we do not stop.

They are innocent,
But thirty die each second,
Because of our race.

We have killed many;
The dodo; quagga zebra,
The list continues.

This will affect us;
It lowers our food sources,
Meaning we could die.

But, you can change this.
Help us stop the countless deaths,
And save the future.

In total, 99% of species that have ever lived on Earth no longer exist. It is believed there have been five big mass extinctions in Earth's history (the fifth was 65 million years ago and was responsible for wiping out the dinosaurs). The cause of each is thought to have involved a significant climate change and some think we're in the sixth right now. Approximately 15,000 new species are discovered each year.

The giant
The great, grand, blue whale,
Echoing its sound wave song,
Echoing its song.

A comforting thought
The ozone layer
Like a jumper covering
Our entire world.

Ozone has the formula O_3. When found in the stratosphere, it filters out a lot of dangerous UV light.

Seasonal cycles
The chilling winter
A harsh time for animals
Most birds will migrate

RANDOM WALKS IN SCIENCE

Psychu
Kills biology
Knifes physics in the shower
Turns into mother

The golden section
We don't always see
Hidden throughout time and space
Nature's beauty trick

The Fibonacci sequence begins as follows: 1, 1, 2, 3, 5, 8, 13, 21, 34, 55, 89, 144, 233, 377 ... and something incredibly fascinating lies within this sequence. If you look at the ratio of successive pairs of Fibonacci numbers, placing the larger number first (e.g. 144/89 and 233/144), as we get further along the sequence the ratio between the two gets closer and closer to 1.618 ... or, more precisely, $\frac{1}{2}(1+\sqrt{5})$. This is commonly known as the golden ratio, φ, or golden section. The golden ratio is often cited as an aesthetically pleasing proportion, and has been used extensively in the arts during the last 2,500 years. Much more interesting than their use by humans, however, is the number of times the Fibonacci sequence and golden ratio appear in nature. If you look at the head of a sunflower, the florets of a cauliflower, the bumps on a pineapple, or the scales of a pine cone, you will notice a particular spiral pattern – so many in one direction and a different number in the other. Nearly always, these two numbers will be consecutive Fibonacci numbers, which as we saw above are the best whole-number approximations of φ.

Infinity
This will keep going ...
On and on and on and on ...
And will never end ...

In mathematics, infinity is a concept rather than
a number, denoting something that is beyond
any fixed bound and that can't be resolved by
counting or measurement, even in theory.

How to read a graph
'Walk before you fly',
This helps you to read your graphs:
It is X and Y.

When plotting a point on a graph, such
as (4,3), it's important to know that the 4
refers to the length along the horizontal
x-axis and the 3 the length up the vertical
y-axis. An alternative version of the above
is 'along the corridor, up the stairs'.

8.17pm, 20 July 1969
World stood still, waiting
First footprint, flag standing proud
One small step for man.

Dr ?
He has a blue box –
It's bigger on the inside.
No one knows his name

A serious charge
Edison was mean –
He stole many ideas.
Tesla was better.

> There is plenty
> of debate on
> whether the above
> is true or not.

Genius
Albert Einstein is ...
A wise guy with crazy hair.
And a formula.

Albert Einstein is responsible for what is probably the most famous equation of all time, $E = mc^2$, where E is energy, m is mass and c is the speed of light. It tells us that mass can be converted into energy and vice versa.

Spontaneous combustion
This is when things burn.
Blowing up out of the blue.
And melt or vanish ...

One Year 8 class was particularly fascinated by spontaneous combustion. One of the most plausible theories regarding its cause, the wick effect (where a victim's clothes soak up their fat, acting like a candle's wick), suggests that the term 'spontaneous' is very misleading.

Don't try this at home
In Science we learned
Cats explode if you feed them
Dynamite and bricks.

> Do you need bricks for the second line to be
> true? Strangely, the lesson in which this was
> jokingly mentioned (you had to be there)
> proved to be one they actually enjoyed.

Our challenge to you
Suck out this water,
Use two straws. One must remain
Out of the liquid,

But still in your mouth.
Pressure diff'rence will fail you.
Now try it at home.

> Believe it or not, if you try to suck up a drink with two
> straws in your mouth but with one of them outside the
> liquid and not in the glass, you'll never succeed ... unless
> you know the secret. You need to somehow block the
> straw out of the glass, and the easiest way to do this is
> to use your lip or tongue. Once you crack this, challenge
> your friends to a two-straw drinking competition and
> make them wonder how you can do it and they can't.

Busy day
I fractured my arm.
I had a greenstick fracture.
I had an X-ray.

Boiling the kettle
Steam is a vapour,
Water can evaporate,
Vapour can condense.

Little fluffy ...
Clouds up in the sky,
Made up of water vapour,
Rain comes trickling down.

Precipitation
A scatter of white,
The backward snowstorm whirls with
Silver clouds adrift.

20,000 feet
Raindrops from the cloud,
Three minutes to hit the ground,
Making a splash sound.

Seismic activity
Ground begins to shake
Tectonic plates fight for space
Worlds collapse above.

Earthquake magnitude is typically a measurement of ground motion, which is then expressed as a value on the Richter scale. It is based on powers of ten, meaning that an earthquake measuring 5 on the Richter scale has a recorded seismograph amplitude ten times greater than one measuring 4.

SCIENCE AT CSG

Science at CSG
The young Camden girl
Stares through her misty goggles
And observes the flames.

Switching on
A dull group of kids
Light with energy and spark
When doing science

Science humour
Chemistry is fun
Due to my teacher's bad jokes.
Sorry Mr Flynn.

If the above is true, it's only
because, in chemistry, all the best
jokes Argon. Okay, I'll get my coat.

Ms Perry
A science teacher;
I like one who's fun and nice
And that's what she is.

My science teacher Ms Keilthy (1)
Science is OK.
But homework equals much fear.
She will kill me fast.

My science teacher Ms Keilthy (2)
The blonde Irish one
Seeks her revenge on me now
For my late homework

Hmmm ... it looks like Ms Keilthy
might be developing a reputation.

What the immediate future holds
Finding out we have
Science next yes yes yes yes
Oh no my homework!

Just for the record, the above
does not refer to Ms Keilthy.

Controlled assessment
An ISA paper
Sucks the joy out of science
A lifeless exam

An ISA (Investigative Skills Assignment) is the controlled assessment part of the exam board AQA's science GCSEs and is worth 25% of a GCSE. Needless to say, it isn't everyone's cup of tea.

Exam stress

Sitting in the hall
Science test in front of me
Will I pass or fail?

Clock is ticking loud
Ten minutes to go. Help me!
But still not finished

My pencil just broke
Halfway through my table! Oops!
What should I do now?

My friends are all done.
Their pens held still in their hands
The last one standing

People leave the room
The last exam is over
Time to celebrate!

Who can't sympathise with the above?
Oh, and it was written after an ISA exam.

Final thoughts
I was told to write
A haiku about science
So I did just that

CONTRIBUTORS

The haiku in this book were written by over 150 students, most of whom were aged between eleven and fourteen.

Scikulogical barriers	Chhaya Lad
Science	Anonymous
The point of science	Anna Ellis
A never-ending story	Edie Jones
What is it good for?	Anonymous
Science vs. religion	Eliot Lambert
Neutral science	Elika Charlton
Have faith	Morgan Conway
Faith in science	Jack Sorapure
Science kills the mood	Claudia Vulliamy
What is a hypothesis?	Muna Abdi
From geek to chic	Anonymous
Triple science	Evie Elliott
Funny fishy science	Scarlett Hayes and Yasmin Omoregie
The element of surprise	Scarlett Hayes and Yasmin Omoregie
He-He-He	Saima Mahamoud
Making a meal of it ...	Marie Meyer
Biological bash	Anna Soroko
Creation	Jessie Crowther

Conundrum	Dolly Banks-Baddiel
In the beginning ...	Elsa Pearl
Night sky	Zada Dzihan, Anonymous, Rashida Hoque
Seirios	Anonymous
Astronomical art	Rachel Moore
Stars	Zakiyah Faruque
Starry Potter	Umi Pawlyn
Danger in space	Emily Copley
The telescope	Anonymous
Shooting stars	Lulu Owen
Gravity in our solar system	Samina Rahman
Copernicus	Jasmine Cheung
Sorry Aristotle	Anonymous
The Earth's orbit	Alex Hill
24:7	Sula Harding-Cornish
Burning bright	Lira Morina
Solar radiation	Anonymous
Here comes the Sun	Zakiyah Faruque
Oblivion	Anonymous
The Goldilocks enigma	Shanjida Rahman
Solar power	Mimi Hitchin
Big brother	Rachel Moore
A hymn to Selene	Rashida Hoque
Planet mnemonic	Rashida Hoque
Venus	Anonymous
Red in tooth and claw	Anonymous

CONTRIBUTORS

King of the gods	Mya Esdale
Satellites galore	Lulu Owen
The last four planets	Henna Von Däniken
A dwindling universe	Tasmin Ahmed
Calling all planets	Bess Gorman
Seventeen syllables	Rosie Harrison-Nirawan
The three ages of woman	Aysha Begum
The splendour of life	Anonymous
Biology bake-off	Poppy Lashley
States of mind not matter	Philomena Wills
Being civilised	Elona Buzhala
Grey matter	Erica Munyako-Trento
Cell-u-like	Sula Harding-Cornish
Not for eating	Meg Bailey
Let's split up	Tahmina Akhter
Cell specialisation	Elisa Ly
White blood cells	Masuma Chowdhury and Shannon Frederick
Immunity	Anonymous
Inoculation	Jess King
Going viral	Anonymous
Triple threat	Anonymous
War games	Anonymous
White blood cells – my little soldiers	Izzie and Hannah
Get your facts right!	Anonymous
Friend or foe?	Layla McArthur-Brown
P stands for phosphorus	Henna Von Däniken

Soilsnake	Stevie-Leigh Doran
The fertility hormones	Hermione Hitchin and Lily Grasso
Family planning	Anonymous
Petal power	Ella Brammer
How bees pollinate	Minnie Fawcett-Tang
Fancy a drink?	Leila Levy Gale
The seven conditions of life	Anonymous
Ventilation	Rachael Boyd
Staying alive	Matilda Glynn-Henley
The story of carbs	Shekinah Reuben
The journey of an oxygen molecule	Elle Clark
Oxygen + glucose → water	Faaizah Siddiquey
Show us what you're made of	Shekinah Reuben
Survival instinct	Shekinah Reuben
Food chain	Jessie Godwin
A plain diet	Blythe Mackintosh
Ebola emergency	Chloe Langmuir
Fitness	Marie Meyer
Food groups	Rachael Boyd
A recipe for a long life	Eden Gray
The dangers of smoking	Jasmin Lagab
Class A advice	Naomi Marchbank
Not an E-asy decision	Layla McArthur-Brown
Make no bones about it	Meelina Isayas
Bone of contention	Rachel Moore
The bone connector	Mimi Hitchin

CONTRIBUTORS

A skin full	Anonymous
IQ test	Tasmine Ahmed, Iman Badran, Meriam Bamijjane
Digestion in a nutshell	Savannah Puleston
Break it down	Anonymous
Bowel movement	Eden Maddix Odeniyi
Autumn's gravity	Victoria Stanescu
Earth watch	Betty Thesmann
Particles	Honey Lloyd
States of matter	Demetra Rankine
Fluidity	Martha Stevenson
Ancient elements	Philomena Wills
Everything is awesome!	Elika Charlton
The atom	Anonymous
More than the sum of its parts	Anonymous
Everything under the Sun	Anonymous
Joining atoms	Elika Charlton
Double trouble	Lira Morina
Atomic orbits	Anonymous
Inside the atom	Betty Steiner
Atomic optimism	Emily Davies
Atomic pessimism	Rachel McHugh
Best friends really	Samira Conteh
Homer Simpson's baseball team	Anonymous
Elementary	Bess Gorman
A clear arrangement	Anonymous
Periodic table (1)	Lucy Perry-Wicks

CONTRIBUTORS

pH scale	Dolly Banks-Baddiel
pH universal indicator	Madeleine Piggott and Anonymous
Did you know?	Alex Hill
Common a-salt	Anonymous
A firework's colour	Maya Hawes
Lilac	Elisa Ly
Fire	Anonymous
The fire triangle	Hanna Gawron
Fire warning	Matilda Glynn-Henley
Complete combustion of fossil fuels	Anonymous
Buy yourselves an alarm	Shannon Frederick
Gunpowder plot	Anonymous
Cold to the touch	Anonymous
Controlling a Bunsen burner	Bessie Gorman
Sulfur dioxide	Mia Holdsworth
A literary poison	Naomi Marchbank
Wow!	Rachel McHugh
A noble delight	Cydney Tripley
A noble caution	Yasmin Diggs-Williams
Dissolving confusion	Morgan Conway
Liquid of life	Anonymous
Water, water, everywhere …	Rose Ames-Blackaby
Drink it all in	Saja Warsame
Water	Zada Dzihan
The test for an alkene	Rachel McHugh
Polymerisation	Anonymous

CONTRIBUTORS

Light	Zakiyah Faruque
Forces	Anonymous
Forces in action	Lara Gebauer
Newton's first law	Anonymous
Friction	Poppy Boswell
Resistance is NOT futile	Isla Dowland
Is work difficult?	Meriam Bamijjane
Under pressure	Anonymous
Newton thinking	Anonymous
Gravity	Anonymous
Gravity on Earth	Weza Da Veiga Baptista
Flying high	Anonymous
Watch your step	Nadia Kabir
Flight	Jessie Crowther
Use the force	Lily Lovering
Have a safe flight!	Catherine McCauley
Crying	Mia Harada-Laszlo
A science lesson from Yoda	Mia Gane
It's simple	Matilda Glynn-Henley
Non-fatal attraction	Calla Brennan Marle
North and south	Anonymous
Electricity	Jasmin Van Den Bos
She's electric	Edie Jones and Aysha Begum
Electric circuits	Masuma Chowdhury
Scientific Russian dolls	Isobel Agnew
The end of the road ...	Kajal Aubeeluck

CONTRIBUTORS

A question of chromosomes	Niyat A'Marium and Sarah Thorogood
Battle of the sexes	Amelia Klein
Pseudoscience part 1	Lily Grasso
Pseudoscience part 2	Dina Ksib
Pseudoscience part 3	Anonymous
Heredity	Emily Copley
Allele together now	Rose Ames-Blackaby
Gene genie	Orla McGroarty
Drawing the short straw	Leila Levy Gale
Body imaging	Nellie O'Leary
Protein synthesis	Elijah Bailey
Franken-science	Anonymous
Cloning v1	Suraiya Aktar, Megan Bradbury, Esther Pigney, Erin Beck
Cloning v2	Suraiya Aktar, Megan Bradbury, Esther Pigney, Erin Beck
Lysogeny	Deba Erhaber
Gender bias	Amy King
Proportional representation	Anonymous
Credit where credit is due	Elona Buzhala
Our shared planet	Jessie Crowther
After life	Minnie Fawcett-Tang
650 million years later …	Lira Morina, Fatimah Ahmed
Natural recycling	Sadia Kabir
The carbon cycle: a modern epic	Anonymous

CONTRIBUTORS

Spontaneous combustion	Erica Munyako-Trento
Don't try this at home	Emily Copley
Our challenge to you	Alex Hill
Busy day	Lily Lovering
Boiling the kettle	Isla Dowland
Little fluffy …	Fatimah Ahmed
Precipitation	Madeleine Piggott
20,000 feet	Isis Hurley-Jones
Seismic activity	Lira Morina
Science at CSG	Salma Mahamoud
Switching on	Hannah Morris
Science humour	Jake Beaton-Rekkers
Ms Perry	Amal Ali
My science teacher Ms Keilthy (1)	Hannah McAndrew
My science teacher Ms Keilthy (2)	Hannah McAndrew
What the immediate future holds	Rachel Mavango
Controlled assessment	Saima Mahamoud
Exam stress	Mimi Torday
Final thoughts	Dolly Banks-Baddiel

ACKNOWLEDGEMENTS

A project like this has required the help of a large number of people.

Without the support and enthusiasm of the school's headteacher, Elizabeth Kitcatt, this book would never have got off the ground.

The heads of English (Angie Fernside) and Science (Martha Hunt and Lynda Charlesworth), along with our colleagues in those departments, worked tirelessly in supporting students at the school to extend themselves in what was an unusual task.

Our editorial team, made up at various points of Rose Ames-Blackaby, Cecilia Bax, Alix Benson, Bluebelle Carroll, Aiyshah Faruque, Minnie Fawcett-Tang, Mia Gane, Amelia Klein, Naomi Marchbank, Mimi Torday, Cydney Tripley and Lily Waxman, have been committed and forthright when discussing all aspects of the book, including layout, cover design and choice of font. It has been a real pleasure watching them immerse themselves in the numerous stages involved in producing a book. Elisa Ly contributed excellent explanations for a handful of the haiku.

Icon Books has been a wonderful partner, with members of staff giving up their time to discuss various aspects of the publishing process. Duncan Heath is a perfectly calm editor, Marie Doherty a brilliant typesetter, and Richard Green a superb cover designer. Hello to Jason Isaacs.

Finally, we would like to thank all the students of The Camden School for Girls, especially those who submitted fantastic haiku for which sadly there wasn't enough space in the book. Without everyone's hard work and considerable effort, this compilation would be nothing more than a pipe-dream.